UNICORN

COLOR BY MATH
ADDITION & SUBTRACTION

This book belong to:

5 - blue

8 – light pink

7 - yellow

6 - dark purple

9 – light purple

4 – dark pink

3 - blue   5 – light pink   4 - yellow

6 - dark purple   2 – light purple   1 – dark pink

12 - light blue   14 – light pink   17 - yellow

11 - dark purple   13 – light purple   15 – dark pink

11 - light blue   12 – light pink   14 - yellow

13 - dark purple   17 – light purple   16 – dark pink

13 - light blue   15 – light pink   14 - yellow

16 - dark purple   9 – light purple   11 – dark pink

9 - light blue   7 – light pink   8 - yellow

13 - dark purple   5 – light purple   6 – dark pink

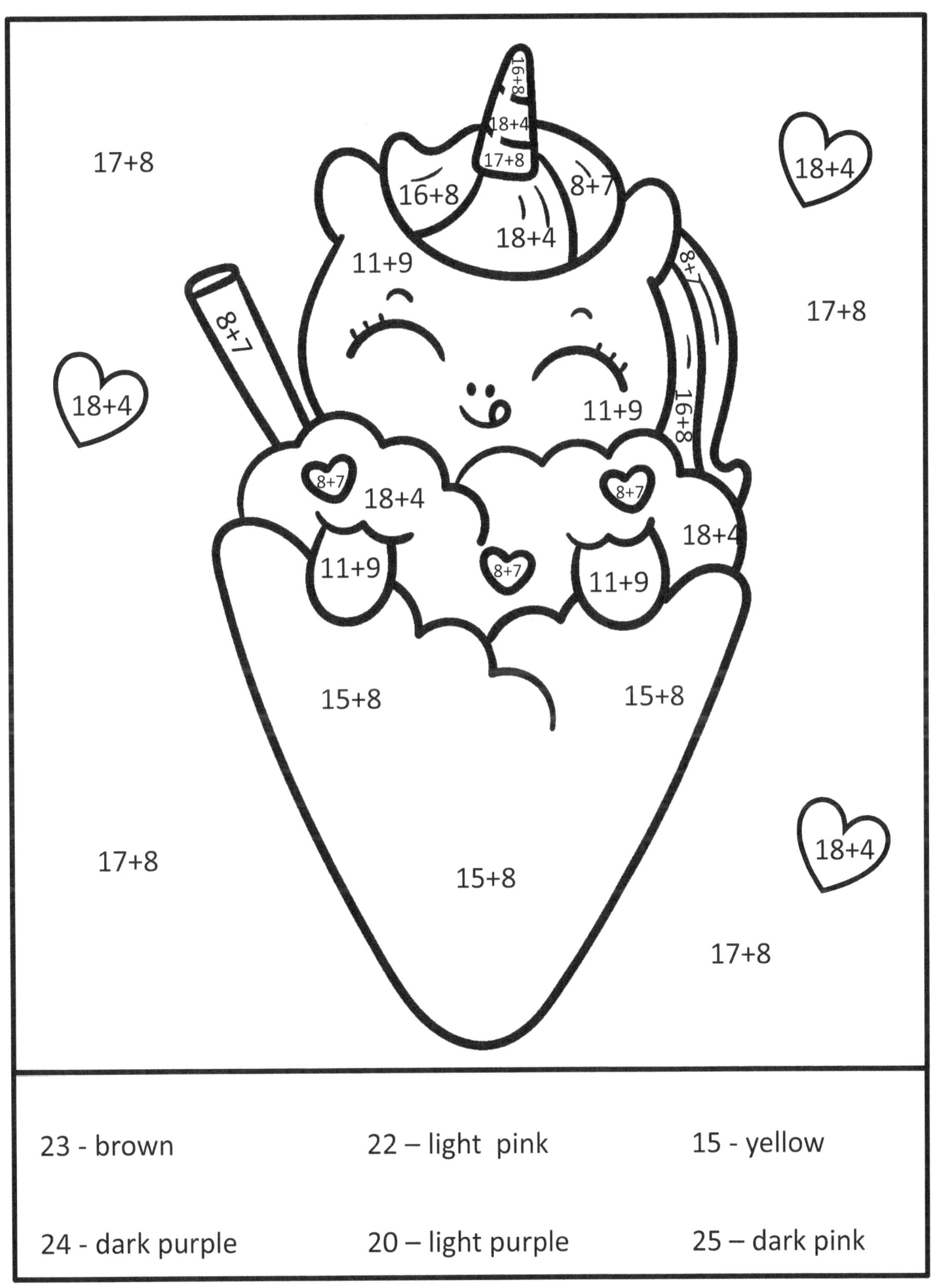

23 - brown   22 – light pink   15 - yellow

24 - dark purple   20 – light purple   25 – dark pink

15 – light blue   12 – light pink   16 - yellow

18 - dark purple   17 – light purple   20 – dark pink

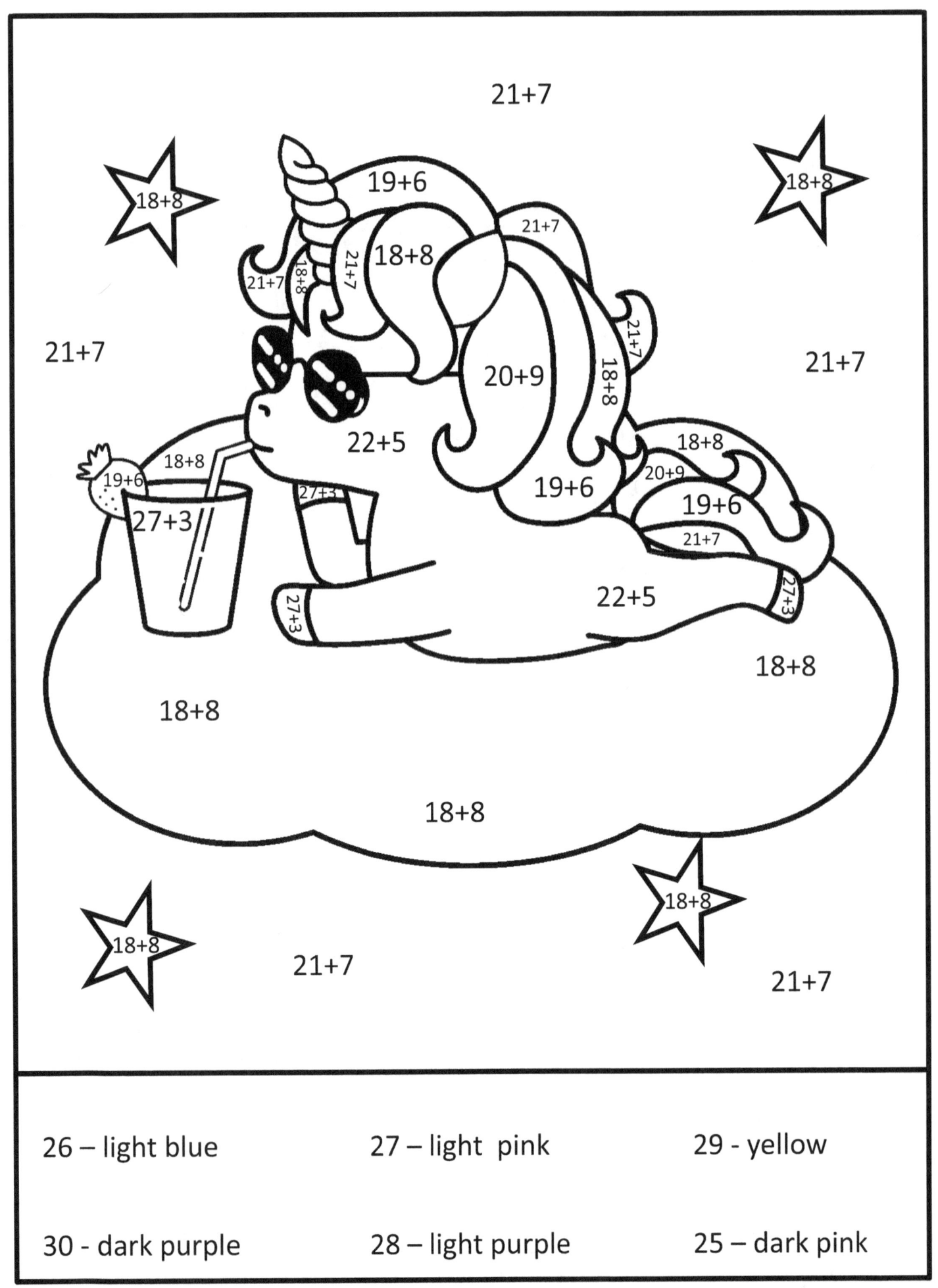

26 – light blue   27 – light pink   29 - yellow

30 - dark purple   28 – light purple   25 – dark pink

26 – light blue   15 – brown   25 - yellow

22 - dark purple   13 – light pink   9 – dark pink

14 – light blue   19 – light purple   20 - yellow

12 - dark purple   22 – light pink   7 – dark pink

18 – light blue   8 – light purple   14 - yellow

17 - dark purple   7 – light pink   6 – dark pink

4 – light blue   6 – light purple   8 - yellow   10 – light pink

12 - dark purple   14 – brown   16 – dark pink   18 – light green

15 – light blue   17 – light purple   4 - yellow   2 – light pink

22 - dark purple   1 – light green   8 – dark pink

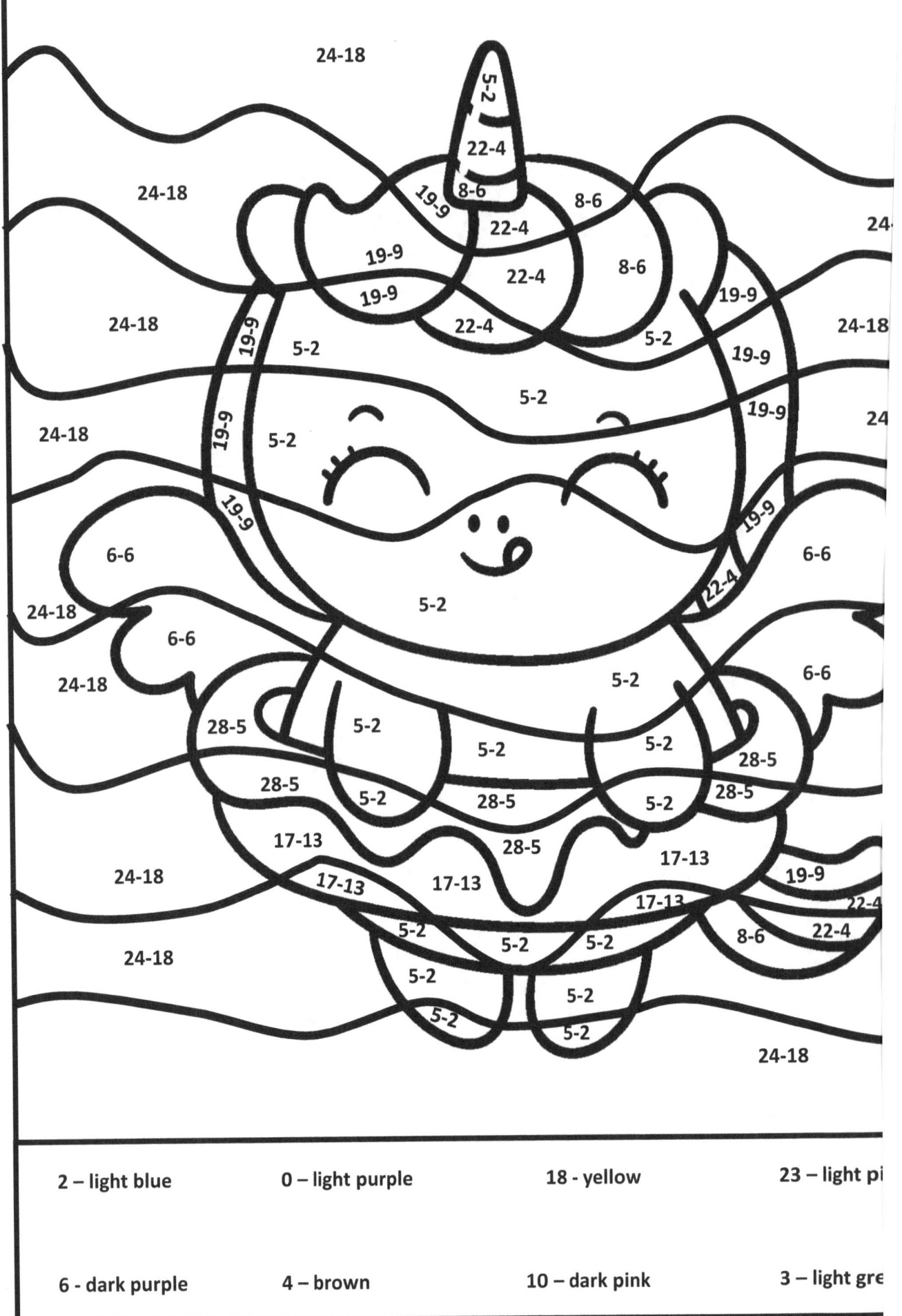

2 – light blue   0 – light purple   18 - yellow   23 – light pi

6 - dark purple   4 – brown   10 – dark pink   3 – light gre

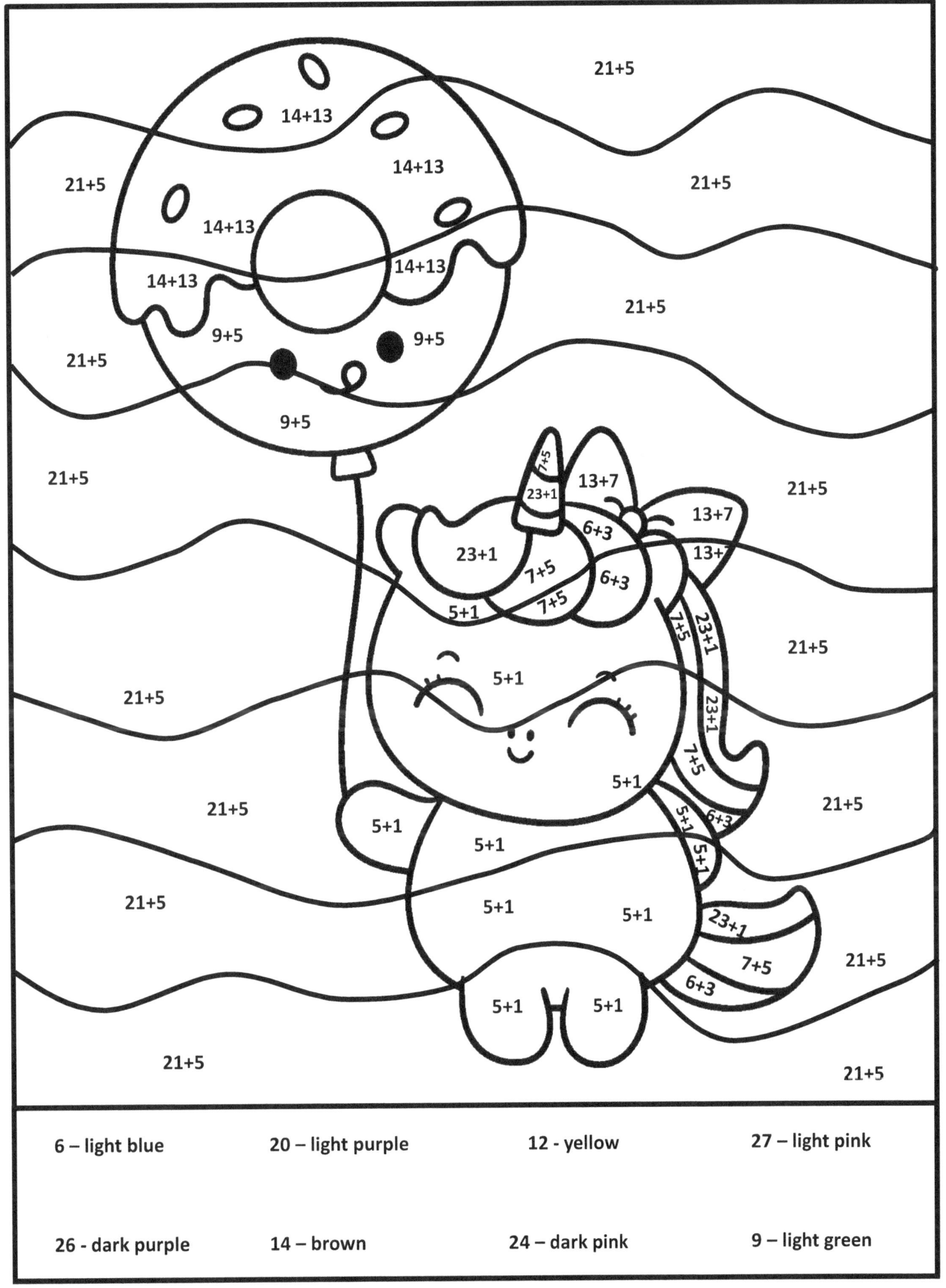

9 – light blue   5 – light purple   10 - yellow   14 – light pink

19 - dark purple   21 – brown   25 – dark pink   28 – light green

1 – light blue   3 – light purple   11 - yellow   14 – light pink

16 - dark purple   25 – orange   20 – dark pink

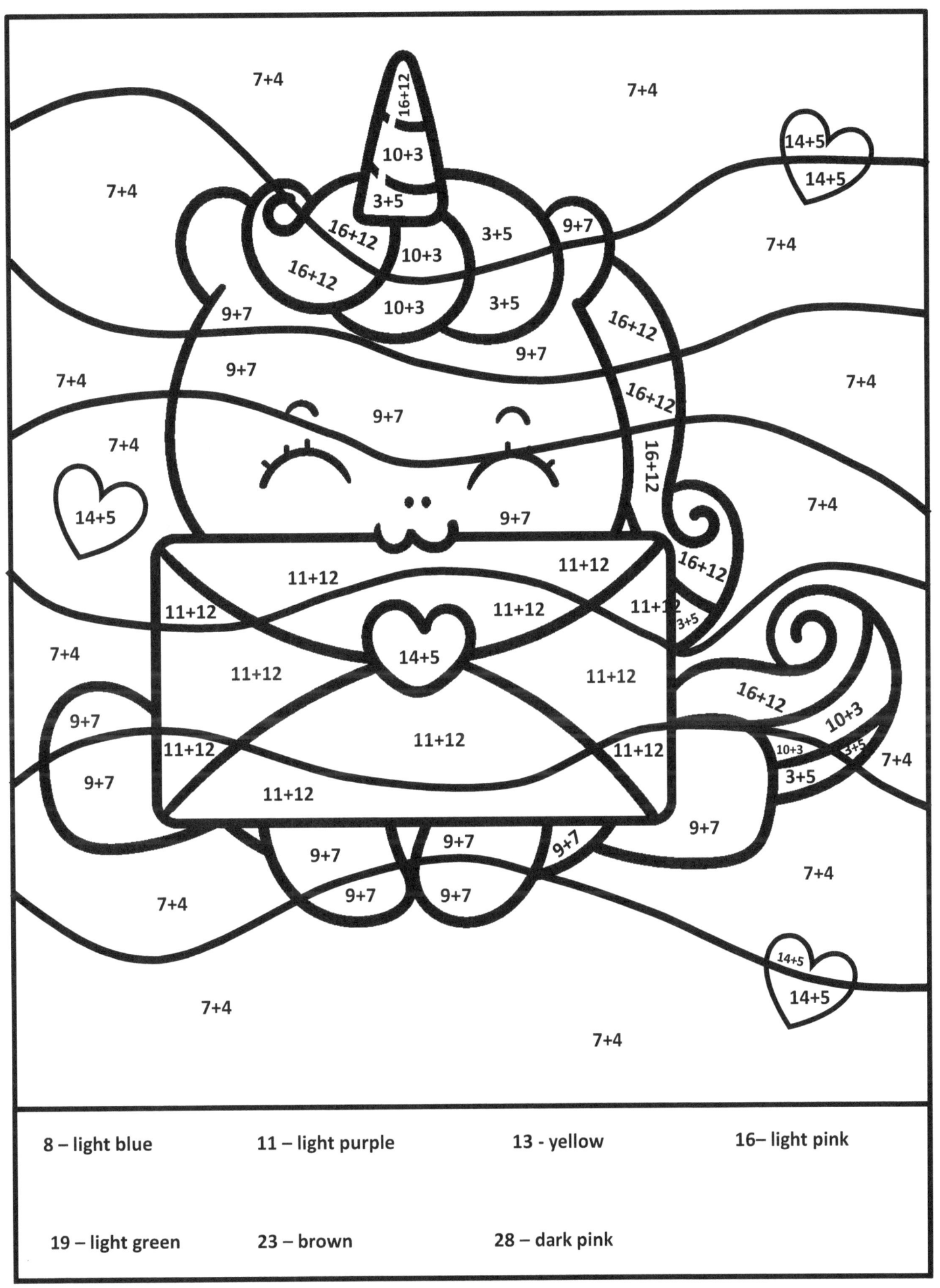

8 – light blue   11 – light purple   13 - yellow   16– light pink

19 – light green   23 – brown   28 – dark pink

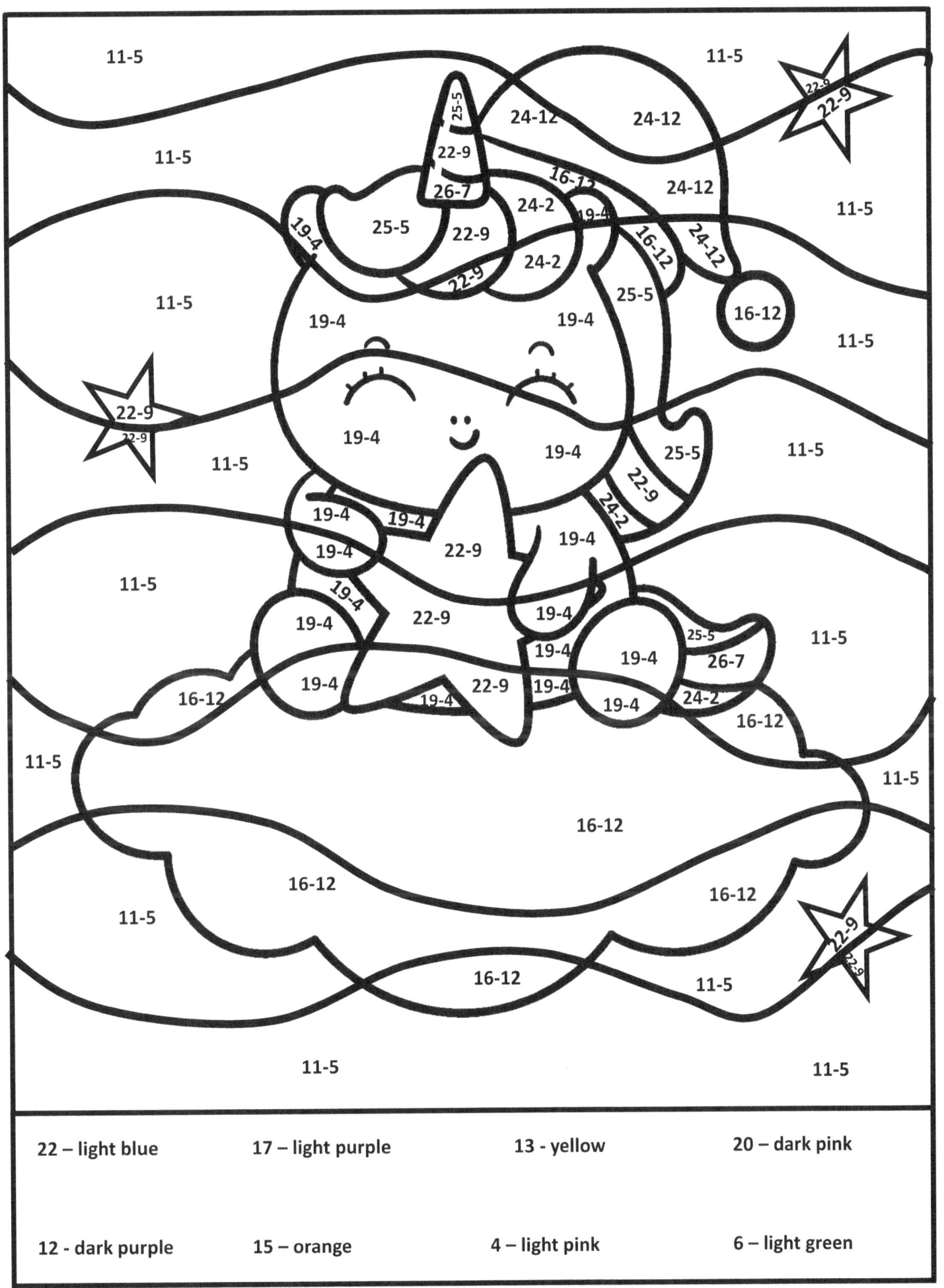

22 – light blue   17 – light purple   13 - yellow   20 – dark pink

12 - dark purple   15 – orange   4 – light pink   6 – light green

8 – light blue   10 – light purple   12 - yellow   16 – dark pink

19 - dark purple   22 – orange   25 – light pink   29 – light green

14 – light blue   12 – o28-11ge   10 - yellow   8 – light pink

15 – brown   29 – light purple   10 – dark pink   9 – light green